THE SCIENCE BEHIND NATURAL PHENOMENA

THE SCIENCE BEHIND NATURAL PHENOMENA

AUGUST RAINES

CONTENTS

Introduction to Natural Phenomena

From hurricanes, waves, and tornadoes to ant colonies, clouds, and the aurora borealis, there is an endless number of breathtaking natural events that have been fascinating and amazing. Some of these phenomena are so beautiful that many people like to photograph them and capture their photographic images. From these examples, it is clear that natural events are an important part of science. The question is: what exactly are natural phenomena? Many people love learning about natural phenomena, but few can define them. That's why we're going to take a closer look at them here. By explaining the term's meaning and discussing whether some phenomena could be considered 'supernatural,' we will take a closer look at it. Do you want to learn more? Let's get this party started!

Nature plays a crucial role in the lives of all organisms, from the smallest insect on Earth to the biggest mammal that roams the world. Yet very few individuals know the concept of a phenomenon. Natural phenomena, also known as organic events, have been defined as the occurrence of unknown incidents. In all of these instances, nature or Earth's natural processes are linked. There are few or no equivalents to most natural events.

Defining Natural Phenomena

As we begin our exploration of natural phenomena, we could surely spend many pages in a dictionary, thesaurus, or animal encyclopedia listing the names of all the things we might consider "phenomena" in the natural world. However, for our current purposes, we need one primary definition: a natural phenomenon is a fact or event that can be perceived by the senses - seen, heard, tasted, touched, or smelled. This definition allows us to see that natural phenomena include everything from the evening star in the sky, to the sound of running water, to the feel of sunshine on one's upturned face, to the smell of a rain shower right after it falls on the warm earth (a smell that scientists have now accurately explained as an airborne bacteria, by the way).

A natural phenomenon can be momentary, like a shooting star, or it can be more enduring, like the Earth's water cycles. It can be local, as when a family from Milan, Italy experiences an earthquake near their home, or global, as when the entire planet experiences the eruption of an Indonesian volcano that darkens the sky and alters the climate. Some phenomena require intelligence and prolonged investigations to observe, study, and make sense of, like the structure of the crust of the Earth and how it shifts beneath our feet. Others rely only on "common sense," what one can see for oneself, irrespective of level of education or social class, like the wonders of the heavens from one's back porch.

CHAPTER 2

Physical Processes in Nature

When fishes dwell in clearer waters, it shall darken the ocean. – William Shakespeare, The Tempest.

If we try to take a step back and look at the concept of "phenomenon" in its most general sense, we might then also wonder what "natural" really means. What is the natural counterpart to the unnatural world? Rearing on ideas discussed in Chapter 6, that of a "world beyond nature that ultimately transcends it." If philosophical interests are out of place, we can at the very least concede that the core of this book has empirical ambitions and illustrates many different technical approaches and interpretations that researchers might want to explore further.

It is particularly easy to focus on the physics of physical processes in nature. The second section moves out into the cosmos to talk about different kinds of stars and whether or not they house solar systems like our own. The third move inwards, examining the forces that have the power to tear the Earth apart and create such terrible phenomena as blizzards, earthquakes, hurricanes, tsunamis, and volcanoes. An empirical outgrowth of our understanding of these physical processes in nature is, of course, to figure out what these physical processes can do with the states that exist within them. This involves looking at the physics of trees, rainstorms, tornadoes, and mountain chains. In all of these cases, gravity, gas or liquid dynamics, heat flow, radiation, and electricity/lights/fires will play significant roles.

Reflection and Refraction of Light

When sunlight falls on a glass slab, it undergoes a slight change in its path, even though the entire glass piece is transparent. This change is natural and occurs when light passes through different transparent mediums, such as from air into water, or from water into oil and innumerable other such combinations. The physical processes behind the incident of light bouncing off a surface so that human eyes can look at objects outside, or light's path bending upon entering and exiting a medium, are known as reflection and refraction of light respectively.

As light approaches the boundary of a reflective medium, the incident wave starts inducing oscillations in the medium molecules closest to it. In turn, these medium molecules induce the oscillations in the ones adjacent to them, and this energy is propagated in the form of secondary spherical waves, known as wavelets. Some seconds after the reflection occurs, all the wavelets superimpose constructively to form the reflected wave. However, in the case of refraction, the energy of the incident wave - which is in the form of a plane wave front - is transformed into a series of wavelets sooner than in the case of reflection. In such a scenario, the refracted wave's plane wave front generally changes due to the difference of density, or hence the speed of light, between the two mediums, thereby bending the light's path slightly. Given that such two optical processes - of reflection and refraction - are healthy, natural, and benign, they produce no harmful effects on the eye.

CHAPTER 3

Meteorological Phenomena

Meteorological phenomena, sometimes simply referred to as weather events, refer to natural events taking place immediately above the Earth's surface. Some of these events are awe-inspiring and even deadly, but all are beautiful and fascinating in their own right. Some may be more common, others more rare, but they are all driven by scientifically comprehensible principles. This chapter moves somewhat away from the book's primary focus and will instead help the reader better understand this genre of appealing natural phenomena.

The scientific principles underlying weather are mirrors of those that govern earthquakes and volcanoes. Meteorology, meteorological sciences, or environmental sciences allow us to predict the possibility and extent of such future events. The following are some meteorological events: 1) Anvil: A type of cloud, usually cumulonimbus, characterized by a dense, curling upper part often resembling an anvil. 2) Aurora: A display in high-latitude regions of the Earth's atmosphere in the form of luminous bands, streamers, or the like; the auroras are caused by the emission of light from atoms excited by energetic electrons that are accelerated in the Earth's magnetic field and collide with the atmosphere.

Rainbows and Halos

Rainbows are one of nature's most splendid effects, created when light undergoes scattering, reflection, and refraction in raindrops. These droplets act as tiny prisms, dispersing each color of light to the observer's eye. Notable for their regular and arch-like luminous spectrum, rainbows are capable of displaying vivid double colors produced by the primary rainbow, in addition to white glare and a dim secondary bow. The inner arc of the bow always exhibits a reddish color, while the outer arc adopts a bluish hue. According to the angle at which the glance is released from the droplet, separate orders of rainbows can emerge. Rainbows can thus form as primary bows (produced by a single internal reflection of light) or secondary bows (arising from two separate reflections). Over the years, individuals around the world have continued to speculate upon and investigate the fleeting natural wonder of rainbows.

Circumstance rarely aligns just right to create the breathtaking optical illusion of a halo in the sky. Shining around the sun or moon, a halo is composed of lightly colored or white rings with beautiful highlights. The most common is the 22-degree halo, which is almost perfectly circular and circles the celestial luminary with a diameter of approximately 46 suns offset. Beyond the 22-degree halo lie the Pollenkress and Lowitz arcs, also sun-oriented, along with the tangential and supralateral arcs. If the outnumbered hexagonal plate ice crystals are spun around in thick cirrostratus or cirrus clouds at high altitudes, halos are formed. The differing orientation and deviation of light give rise to the different sorts of halo.

CHAPTER 4

Astronomical Phenomena

Solar and lunar eclipses: The most awe-inspiring astronomical phenomenon is a solar and lunar eclipse. When the moon's shadow falls on the Earth, visible as both a partial and total solar eclipse, the world seems to blank out. Eclipses occur because the moon revolves around the Earth at just the right angle to fall across the face of the sun. On the other hand, a lunar eclipse occurs when the Earth's shadow falls on the moon. The sun and the moon are opposite each other at this time, causing the moon to appear red, or blood-tinged. The first red light is refracted through Earth's atmosphere, causing the moon to appear red. A lunar eclipse is never total and will never last longer than one full night.

Waterspouts: Waterspouts are tornadoes that form over water. Some can be destructive and dangerous, while others are weak and short-lived. Waterspouts are often seen in warmer parts of the world or at subtropical latitudes. They occur along the edge of a thunderstorm and are widespread during hurricanes. Waterspouts are believed to be rendered from the supercell that produces a tornado. Both tornadic and fair-weather waterspouts are resistant to radar detection. Dedicated waterspout research exists, but over land, not over bodies of water. Waterspouts can overturn small boats, and their strong winds and waterspouts have thrown fish from the water to land. Waterspouts are not as violent as land tornadoes; the rotation is usually clockwise in the more deadly waterspout. Waterspouts are capable of moving onshore, killing people and causing damage. Tornadic waterspouts are usually temporarily classified. No scale to rate or rank waterspouts currently exists.

Why Is the Sky Blue?
Why is the sky blue: Optical explanation
The mesmerizingly deep blue sky is a sight that most people universally associate with good weather. Yet, many people don't actually know why the sky is blue, aside from the fact that light is being scattered around in the Earth's atmosphere.

To begin, the sun emits a type of light that is white in that it is a combination of all of the colors within the visible spectrum. This means that when sunlight shines through a prism, it produces a rainbow, which is a visible separation into all of the different individual colors of light.

When this light penetrates the atmosphere and moves throughout it, each of the different wavelengths of light interacts differently with its respective surroundings. As Sir Joseph Norman Lockyer proposed before he passed away, the nitrogen and oxygen molecules in the Earth's atmosphere scatter the blue light waves of the visible spectrum, yielding the blue sky that we observe.

The violet and green light waves (with the violet having the shortest wavelength and the green, the second-longest) also become scattered to a lesser effect, while the majority of the yellow, orange, and red light waves continue on in their customary straight paths as they penetrate the Earth's atmosphere.

The fact that the color less affected by this scattering of light is yellow allows for us to come up with an explanation to the question as to why sunsets appear red, a phenomenon known as Rayleigh scattering. As the sun sets, the light enters the atmosphere at an angle, and that allows for the blue to have more time to scatter and be exhausted by the time it reaches our eyes. So, the color we recognize the most is the red scattered light waves that have had the most time to deviate to the troposphere of the Earth's atmosphere, giving off the appearance of a red sunset.

The Northern Lights

We often marvel at the different natural phenomena in our world. Some of these phenomena are a result of astronomical processes. For instance, a total solar eclipse, wherein the moon blocks out the sun, can be viewed when the orbital paths of the sun, moon, and Earth align. Binoculars are a good tool for observing the faint outer layers of the sun and the solar corona during this event. Another phenomenon off our coast exists on the bottom of the ocean. Brown algae, or kelp, possess air bladders that become filled with gas, enabling them to float on ocean currents. Ocean currents transport these algae to our coastline, where they often wash up as a result of strong surf. This process reinforces the notion of the interconnectedness of our planet. There is a third natural phenomenon, however, that lights up our night skies. On a clear, dark night, people at high latitudes may have the opportunity to view this ethereal light show. Known to the scientific community as the auroras, it is more commonly referred to as the Northern Lights (or Aurora Borealis at the North Pole) and Southern Lights (or Aurora Australis at the South Pole).

If asked, some people might explain the auroras as colored light reflections off the icy tops of the Earth's atmosphere. While icy molecules do play a part in the eventual colors experienced near the surface of the planet, this is not the process that generates the aurora itself. In the sun's corona, and sometimes as a result of different solar activities, solar flares or coronal mass ejections can eject solar material into space. This results in a solar wind that carries solar plasma and trapped magnetic fields into interplanetary space. The solar wind, which travels at a speed of about 1 to 3 million miles per hour, can take from a few days to a few hours to reach the Earth. Upon arrival, magnetic storms can cause incalculable chaos, such as damage to electric grids, radio blackouts, and disruption of satellite communications. Prior to this apocalypse, however, plasma will compress the magnetic field, distorting the shield while also squeezing and stretching it. In the meantime, plasma begins to flow around the Earth's magnetic shield and sheet plane. When the plasma particles strike the geomagnetic field, they accelerate into the Earth's protective magnetic field. Following that, these charged particles will move across the planet's magnetic fields to emit photons. Left to their own devices, these particle reactions occur at altitudes of 60 to 150 miles. This process forms what scientists refer to as the aurora borealis, an electromagnetic event of studied beauty that is often displayed to the public.

CHAPTER 5

Geological Phenomena

Behind the Facade: Geological Phenomena

1. Geological phenomena are some of the most captivating subjects of our planet. From tectonic shifts to fascinating minerals, from volcanic eruptions to the eventual carving of beautiful landforms, the processes of nature have always been at play in shaping the landscapes that we can see. As these remarkably dynamic processes continue to shape the surface of the Earth, is it not worth considering the 'hows' and 'whys' behind them? As we scratch just below the surface, we find our understanding of Earth's most captivating tales altered. What hostilities lie between moving tectonic plates? Why do volcanic eruptions occur? How is a cave system formed? And does it really feel different to walk on a bed of red hot coals?

2. Deep beneath our feet, the rocks of the Earth slowly creep in response to heat flowing in convection currents in the mantle. The plates of the Earth's crust float on this, but their uneven speed creates irregularities and tension. When the breaking point is reached, the quake causes movement along faults, and the earthquakes that are reported are magnitudes of energy produced. From this, surface waves are created that are able to be transmitted through air. Hence, as in the bumper of a car, the magnitudes of the energy are mass,

speed, and distance. Coastal earthquakes, in which plates are pushed together, or reverse-thrust faults, often cause larger tidal waves, but that is not to say earthquakes at oceanic faults are safe or hazardous spill over- the power is in the shaking of the Earth.

Volcanic Eruptions

Over large swaths of time, the Earth often appears to slumber, with mostly gentle, virtually imperceptible geographic change. But, with sudden reminders, events such as volcanic eruptions underscore explosive geological forces at work beneath the ancient crust. Responsible for a great deal of short-term damage but also for many longer-term beneficial changes, volcanic eruption is one of the most powerful of our planet's natural phenomena.

The primary driving force enlarging the volume of our planet and shaping its geographic features is the energy derived from deep within its viscera, mostly from nuclear decay. This energy moves tectonic plates, releasing some of the pressure that has built. This volcanic release can last minutes or months, while massive amounts of released gases, rocks, and ash shape our world. Every type of volcano has its own unique hazards associated with it. For instance, cinder cones, commonly found in Central and South America, can tunnel into a volcano's magma chamber and blow rock and ash a dozen kilometers into the air. In Hawaii, on the other hand, lava flows from shield volcanoes move so slowly that people and animals can usually evade the flows, though they will destroy buildings and highways. Additionally, from beneath Antarctica's vast ice fields, even volcanoes can erupt explosively, ejecting molten rock, sulfur, steam, and ash into the air. They produce plumes of mephitic "vog" (volcanic smog) as a final outburst.

CHAPTER 6

Biological Phenomena

We often hear how living organisms possess abilities beyond the scope of human technology, how they perform feats that scientists aspire to replicate or improve upon. Delving into this topic of biological phenomena is much more fascinating when one realizes that the scientific basis for such capabilities is still under investigation today, opening up new doors for discovery daily. This section ventures into the countless possibilities available within the world of living systems, examining four of these phenomena and the aspects of scientific knowledge established therein.

Systems of reproduction showcase an intriguing spectrum of diverse methods fulfilled by various organisms, including plants, animals, and fungi, and involving both sexual and asexual means. Offspring dictated by such reproductive strategies possess their parent's genetics—except for a special few, such as bees, in which a haplodiploid system translates into a complex web of genetic relationships. Innovations offering efficient energy gain, waste removal, and sensitivity have been sought in the design of implantable medical devices such as artificial hearts, retinas, and contact lenses. Yet various types of living organisms—the fungi, plants, and, to some extent, animals—offer potential at least equivalent to, if not greater than, these devices. During the process of development, a living organism gains its form and function through repeated use of a repeating mechanism. The underlying design of this mechanism is the means by which multicellular organisms must generate a highly complex tissue structure.

CHAPTER 7

Bioluminescence

When you stand upon the shoreline at dusk, listening to the waves and feeling the salty spray of the sea on your lips, you are undoubtedly looking for some enlightenment. But it is unlikely that you think about the light shining through the waves in a strictly biological manner. The ocean, with its myriad denizens, has always held a special place in human imagination, and light has always had a mystical significance. Yet, while most marvel at bioluminescent light displays in a vague way, or ascribe their beauty to simple phosphorescence, this spectacular phenomenon has a wealth of science behind it that would deepen a person's appreciation of bioluminescence even while removing some of its magic.

Bioluminescence is one of science's greatest discoveries. Here, several separate biological organisms work together to emit light, and the process is well studied enough to control the intensity and duration of the glow at will. In short, a luciferin (a pigment that's actually been removed from living systems and used to monitor a variety of chemical reactions) is oxidized in concert with the enzyme luciferase, with the resulting excited state decay yielding energy in the form of light. In order to prevent wasteful and dangerous spontaneous oxidation, ATP and molecular oxygen are also used in the process helped along by the energy of the original conversion of luciferin to oxyluciferin.

CHAPTER 8

Chemical Phenomena

While there are powerful things happening on a cosmic scale, there are also some equally cool things happening on a much smaller scale. Say, the scale of molecules and atoms. These small-scale occurrences are often driven by chemistry or the way that substances react with one another. We even have a special area of science that specializes in studying the behavior of extremely tiny things. Quantum mechanics is a branch of physics but has a lot of overlap into chemistry because of how the super tiny world is, and as it turns out - a lot of the behaviors that quantum mechanics explains are actually chemical phenomena. Here are a few other examples of naturally occurring chemical phenomena.

Sometimes an entirely new crystal is born when two liquids are combined. When vegetable oil and saltwater are combined just so, the salt makes the top layer of the oil droplets clump together and form salt. While it is true that nothing is left of the original oil (and one of the best ways to make your salad delicious is to put oil on it), this process of combining is interesting because of what we're left with when it's done. That crystallized top layer on the oil/water mix is weirdly squishy and quite pretty! Some people spend their whole lives breaking all manner of things in order to figure out just how they can fit them back together, and by cracking molecules and figuring out the rules that govern how all 3 million something different molecules can play together, chemists make all new things out of very old ingredients.

Acid-Base Reactions

Here, we unravel the chemical magic that transpires during acid-base reactions. In the world of solutions, substances interact and often transform. Their interactions are explained by views developed over centuries, with each generation adding a little more knowledge to what came before.

In the field of chemistry, solutions provide a very expressive representative of these transformations. Those particular solutions are known as "acids" and "bases," and today we will watch as they intermingle. But even more than the performance of these solutions, we want to understand why they do what they do. Regardless of class, human interaction frequently obeys fairly standard rules of behavior, and today we shall focus on rules of behavior in chemical interactions. Our goal today is to give you a scientific perspective on these well-known actors in the chemical theater of operations. Our specific objective will be to describe what is occurring at the atomic level for the coming together of solutions. For now, know that this phenomenon is a chemical interaction, in many cases a chemical transformation. We can watch in amazement at what can occur in the natural world.

CHAPTER 9

Conclusion and Future Directions

In recounting the scientific discoveries and the natural phenomena observed and explored in this book, we harvested the marvels of the natural world, from the domesticated behaviors of animals, to the origins of the Sun's light, to the intricacy of the snowflake. We chose these topics to study out of our own interests and are conscious of the inevitable blindness that our subjective motives cause. Yet we also are struck by how the objective world seems prepopulated with innately interesting phenomena; there are always aspects of the universe that beg for attention, fixed points of light around which our understanding coalesces, the explanation of which becomes the mark of scientific progress. The deepest reaches of physics certainly have no end, but there is still the sense of a handful of bright stars, a collection of results and questions that is as good a starting point as any for approaching questions about the deepest nature of reality.

While some may still disparage these investigations as overly speculative, we would like to think that the wonder evoked is in itself valuable. A confrontation with the beauty of the universe can provide inspiration for questions and curiosity toward discovery, capturing the imagination of scientists and young students alike. Each of our questions and explorations in this work began by accepting a husky sigh of wonder, a resonance provoked by our surroundings. Only now are we beginning to adopt a more terrestrial language to engage with such beauty. For the challenges to explore that which takes us beyond this world, which await the next generation of scientists, we end our tale with a spirit of curiosity and hope for the future.

www.ingramcontent.com/pod-product-compliance
Lightning Source LLC
Chambersburg PA
CBHW051421130726
47989CB00007B/3022